GERMAN TRUCKS & CARS IN WORLD WAR II VOL. III

OPEL AT WAR

Eckhart Bartels

Left: An Opel Blitz supply column outside the port city of Derna in North Africa in September 1942. The last vehicle in the column is an all-wheel drive Blitz 3-ton, a type favored by Rommel. In the foreground is a 3-ton S-Blitz.

SCHIFFER MILITARY HISTORY
West Chester, PA

Translated from the German by David Johnston.

Library of Congress Catalog Number: 91-60856.

Printed in the United States of America.
ISBN: 0-88740-309-3

This title was originally published
under the title,
Opel im Kriege,
by Podzun-Pallas-Verlag GmbH,
6360 Friedberg 3 (Dorheim).
ISBN: 3-7909-0209-8.

Published by Schiffer Publishing, Ltd.
1469 Morstein Road
West Chester, Pennsylvania 19380
Please write for a free catalog.
This book may be purchased from the publisher.
Please include $2.00 postage.
Try your bookstore first.

Opel cars and trucks were the most widely used of all Wehrmacht vehicles and saw service on every front. Here an Olympia-38 and a 3-ton Blitz are seen on a street in Simferopol in the Crimea in the winter of 1941.

FOREWORD

In contrast to other German manufacturers, Opel never showed a marked or heightened interest in the armaments industry. The attitude of scepticism held by the National-Socialist leaders toward the American management of the Opel factory allowed the firm's directors to remain true to the Opel tenet of allowing the widest possible segment of the population the enjoyment of the automobile by keeping prices low. This was achieved through the rational and economical production of large numbers of technically mature yet simple designs. At the same time, this was also one of the reasons why Opel vehicles were to see service on every front during the Second World War.

For most of the photographs reproduced here we are indebted to war correspondents, who themselves often visited the various theatres of war in Opel vehicles. For making these photographs available special thanks go to the Opel Press Agency as well as the members of the Old-Opel Association, who supported my work with photographs and information.

Eckhart Bartels
Old-Opel Archives

Ronnenberg, April 1983

An Opel Kapitän cabriolet at Cape Aju-Dag on the south coast of the Crimea, northeast of Yalta, in April 1942.

The Opel Blitz

Anyone wishing to describe the use of Opel vehicles during the Second World War would naturally think first of the Blitz truck. No military vehicle was to be encountered on every front in those years as frequently as the Opel Blitz with its variety of body styles. No motor transport Sergeant in the German Wehrmacht would have wanted to do without the dependable, almost indestructible, Opel Blitz, whether it be on the Western Front, in the Balkans, the heat of North Africa, in the muddy period or the cold of Russia, or in the barracks of Germany.

The Opel Blitz owed its legendary dependability to its proven overall concept, to which the use of a drive train and components from automobile production made a significant contribution. In this way Opel could, after making the appropriate modifications for use in trucks, fall back on many proven components. The advantages of the unitized construction system for series production had already been recognized and, following the acquisition of a controlling interest in the company by General Motors of Detroit, was adopted.

1931 saw the beginning of the new era of the Opel truck, which was to end in 1944 with the destruction of the production facilities in Brandenburg on the Havel. Opel was already one of the few German companies to manufacture passenger vehicles and trucks and, during the First World War, had been one of the suppliers of the standard 3-ton truck to the German Army, although only in modest numbers.

Opel continued to build limited numbers of smaller trucks during the 1920s. The success of the Opel Blitz trucks began with the introduction of the 2- and 2.5-ton trucks at the beginning of the 1930s.

The Opel Advertising Department organized a competition to find a suitable name for the new truck. It had to consist of no more than five letters and was to be understandable nationally and internationally. First prize was a 1.2-liter, 22 h.p. Opel Sedan, while the winners of the second to fourth prizes were each to receive a 500-cm3 Opel Motoclub motorcycle from the motorcycle factory in Brand-Erbisdorf in Saxony. It took quite some time to evaluate the astounding 1.5 million entries. It was not until 25 November 1930, on the occasion of a dealers convention in the large *UFA-Palast* in Frankfurt/Main, that sales director P. Andersen was able to announce the new name for the 1931 Opel truck: the OPEL BLITZ.

The principle behind production of the Blitz fast truck was a success from the beginning: low weight combined with a large pay load. The Blitz engine was proven technology from the Buick company. Opel was able to take over the complete production facility for the 6-cylinder, 3.5-liter Maquette engine.

The upswing in German truck production began with the "Opel Blitz." Here drivers in Opel uniforms prepare to deliver a week's production of Opel one-and-three-quarter-ton trucks. (1931)

THE BRANDENBURG FACTORY

There now began a period of steady expansion. It was decided on 1 April 1935 to construct a new truck factory in Brandenburg on the Havel River as the second Opel production location. The new factory was built northwest of the city on 850,000 square meters of land, or 340 Prussian *Morgen* (acres) as Opel, with an eye to its new surroundings, stated in its press release. Built in only 190 working days, it was the most modern truck production facility in the world. The new Opel factory was a model example of a rational production process.

All the manufacturing processes, from raw materials to finished trucks, were united in a giant, well-lit assembly hall of 178 x 136 meters. A track of the German State Railway led directly into the factory and eased the unloading and direct storage of materials. Thirteen parallel machining lanes turned raw materials into crankshafts, camshafts, cylinder blocks, gears, frames and front and rear axles. The parts travelled on 27 completely automatic conveyor belts with a total length of 5 kilometers to the final assembly line. When production began 50 trucks per day were coming off the assembly line in a daily eight-hour shift. By later increasing to three working shifts, output was increased to 150 Opel Blitz trucks per day without difficulty.

This was a good starting point for a type of truck which, in 1936, was to become the first to meet the Reich Transportation Minister's specifications for a production truck capable of cross-country travel.

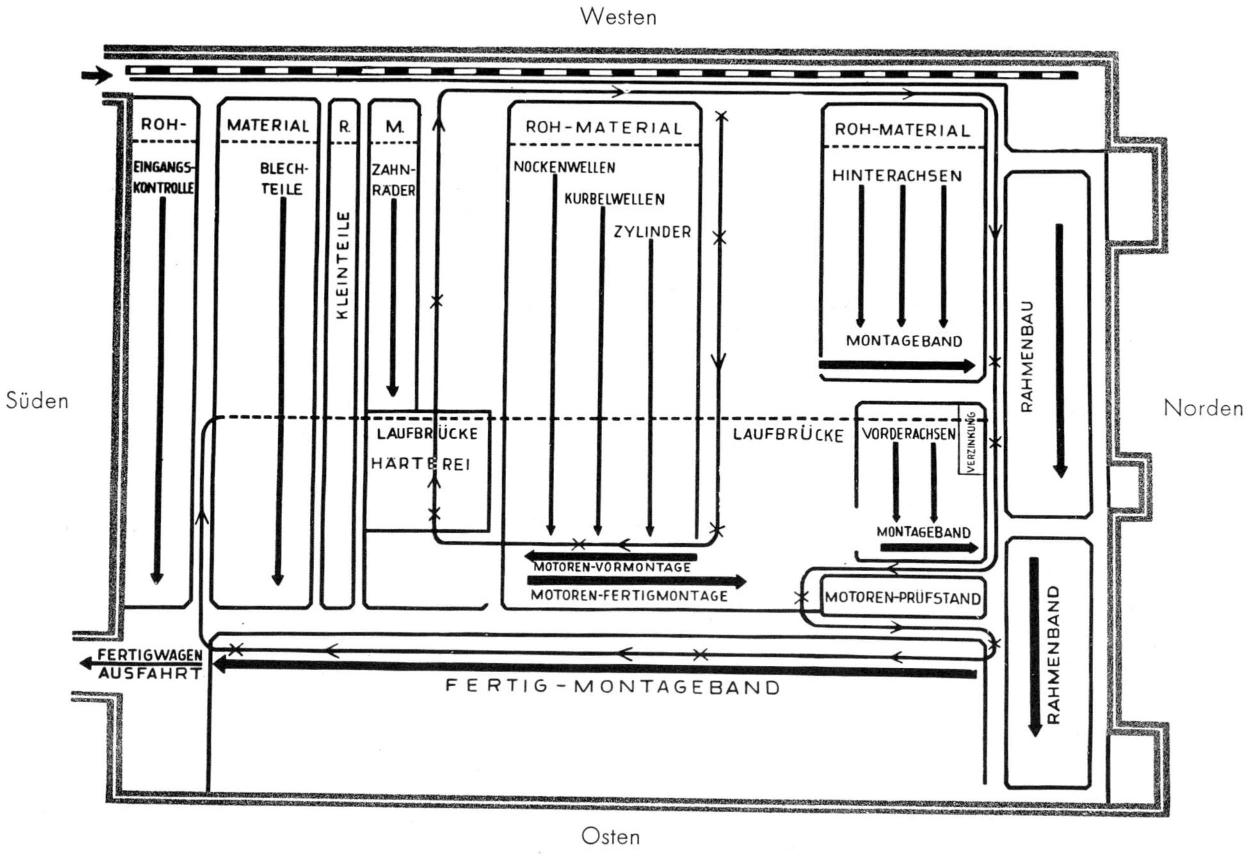

In 1937 the Opel Truck Factory in Brandenburg was a model example of efficient production methods and modern architectural design.

Opel Blitz production in the new Brandenburg factory at the end of 1936.

Left below: A day's production of Opel Blitz trucks await final inspection.

THE "S" TRUCK

The second statute to amend the motor-vehicle tax law, which was introduced on 28 February 1935, included significant easements for owners of commercial vehicles, especially for cross-country capable "production trucks." A year later design specifications for commercial vehicles with limited cross-country capabilities were issued by the Reich Transportation Minister. Among the requirements were:

1. favorable pay load-to-weight ratio
2. favorable power-to-weight ratio (relating to the maximum allowable weight)
3. faultless running of the engine when truck was in a sloping position
4. at least five forward gears with cross-country gear
5. adequate ground clearance
6. limited overhang (greater angle of slope)
7. greater compression and rebound range of spring suspension
8. oversize tires

The first truck acknowledged as "cross-country capable" was the 3-ton Opel Blitz "S" introduced in 1936. It was initially powered by a 3.5-liter, 65 h.p. engine, which was replaced in 1937 by the new 3.6-liter, 75 h.p. engine. The "S" designation may have stood for Standard, Series or perhaps Subsidy, in reference to the standard First World War truck. This Opel vehicle of the new "S"-Class was the first to enjoy a reduced 33.3% tax rate (annual tax was 252 RM instead of 378 RM).

The rest of the German commercial vehicle industry did not follow Opel's example until a year later, when the Büssing-NAG 654 and, in the 3-ton class, the Type 30 Burglöwe were introduced. Beginning in 1937 Hansa-Lloyd, Ford, Henschel and Magirus also offered 3-ton "S"-Trucks. Later arrivals on the market were the MAN Z2 with a 3.5-ton pay load (1938) and the 3-ton Mercedes-Benz LGF 3000 (1939). In the meantime, however, Opel had captured a 35.6 percent share of the market for itself, a head start that none of its competitors could overtake. In the pre-war period the Opel 3-ton "S" proved a success in every industry, just as it did in the expanding Wehrmacht, whether it was on the job at the major construction sites of the *Reichsautobahn*, in quarry work or the construction of the Westwall.

The Opel display at the 1938 IAMA, the International Automobile and Motorcycle Exhibition in Berlin. The entire line of Opel commercial vehicles was on display, from the small P4 delivery van through the small Blitz box-type delivery van to the 3-ton truck, including the military version, and the low-frame passenger bus chassis.

TESTING FOR THE REAL THING

Because of the extensive nature of the fortification of Germany's western frontier, the Inspector General of German Roads, Dr.Ing. Fritz Todt, called on the *Reichskraftwagen-Betriebsverband* (RKB) to participate in the construction of the Westwall. Therefore, in 1938 the RKB took over the transport organization of eleven to twelve-thousand vehicles engaged in construction of the Westwall from private industry. Experience with the many types of trucks involved was advantageous to the Wehrmacht. The Opel Blitz proved itself so well that the Wehrmacht ordered larger numbers of the type.

The testing of suitable German vehicles for the Wehrmacht had begun much earlier, however, with the formation of the NSKK (National Socialist Motor Corps), which carried out trials under the cover of sporting events. Although not designed for them, Opel vehicles were quite successful in many sports races, whether long-distance endurance races or cross-country drives. Special versions of the Opel 6-cylinder car for the NSKK appeared in 1936. In numerous comparison tests these 2-liter and later Super-6 sports two-seaters and special cross-country sports cars demonstrated the usefulness and dependability of the standard production components used in their construction.

Although Opel vehicles had repeatedly emerged successful from winter and mountain trials, the

Winter trials in East Prussia at the end of January 1935. The Opel team fights its way over muddy roads. All of the Opel vehicles which started successfully reached their objectives.

Right: Opel 1.2-liter during the Brandenburg cross-country drive of 1934. It is seen here negotiating a steep bank with a gradient of 50 percent.

The ability of Opel cars in operate in difficult terrain was confirmed in various trials and put to the test in sport races beginning in 1934. Opel even built small numbers of two-seat sport and cross-country sport models based on the Opel 2-liter Type "6" and "Super 6" models.

These special vehicles were featured on the cover of the June 1939 issue of the former "ADAC-Motorwelt" which featured a report on test drives carried out by the National-Socialist Driving Corps (NSKK).

company received no special commission from the Wehrmacht to develop a Kübelwagen or light utility vehicle. The mistrust caused by the company's American management was too great. As a result of this, the only Opel Kübelwagen to enter service were conversions of civilian cabriolets. These open vehicles were based on the Opel 1.2-liter, Opel P4 or Super 6. With the beginning of "vehicle conscription" in September 1939, many Opel Sedans and 1.3-liter commercial vehicles were also converted. One vehicle in great demand was the large Opel Admiral 3.6-liter model, on account of its load-carrying ability and interchangeability of engine parts with the Opel Blitz truck.

Opel had thus already complied with the demands of the Plenipotentiary of Motoring, General Staff *Oberst* Adolf von Schell, whose objective was the standardization and rationalization of the German automobile industry: fewer models in greater numbers and interchangeability of parts. Although the regulations known as the "Schell Plan" were first issued in March 1939, Opel had already largely achieved standardization of components within its vehicle programs by 1938.

The Opel car soon had to prove its "military" suitability as well. Driving a 3.5-liter Blitz, Franz Traiser of Rüsselheim won a gold medal in the "Three-Day Harz Drive" which was held from 9-11 May 1934.

Vereinheitlichung innerhalb des Opel-Fahrzeug-Programms. Über die gleichartigen Ausführungen von Pkw und Lieferwagen bei Kadett (0,5 t) und Olympia (1 t) hinaus sind gleich:

Motor	Super 6, Kapitän, 1,5-t-Lkw-Blitz. Admiral und 3-t-Lkw-Blitz.
Kupplung	Olympia, 1-t-Lieferwagen, Super 6, 1,5-t-Lkw-Blitz. Admiral und 3-t-Lkw-Blitz.
Getriebe	Olympia, Lieferwagen 0,5 und 1 t. Super 6 und Admiral. Verwandt 1,5-t-Lkw-Blitz.
Hinterachse	Super 6 und 1-t-Lieferwagen. Admiral und 1,5-t-Lkw-Blitz.
Bremsen	Kadett, Olympia, 1,5-t-Lieferwagen. Größere Bremsbacken: Super 6 und 1-t-Lieferwagen. Hauptbremszylinder und Handbremse gleich: Admiral und 1,5-t-Lkw-Blitz.
Lenkung	Kadett, 0,5-t-Lieferwagen, Olympia, 1-t-Lieferwagen. Super 6, 1,5-t-Lkw-Blitz, Admiral.
Türen	Kadett und Olympia. 1-t-Lieferwagen, 1,5-t- und 3-t-Lkw.

OPEL CARS AT WAR

Although no Opel cars were built specifically for the Wehrmacht, Opel sedans and cabriolets were to be seen everywhere in military service. The majority of these were vehicles which had been "conscripted" at home or confiscated in occupied territories. The Opel was also a popular car outside Germany, 50% of production being exported. There were factors other than availability which resulted in so many Opel cars being used by the Wehrmacht: its conventional rear-wheel drive, its easy-starting engine and its universally praised reliability helped German soldiers out of many mud holes, across impassable plowed fields and through many cold Russian nights.

A few units used Kübelwagen which had been converted from 1.2-liter or P4 commercial vehicles. The suitability of the robust chassis was enhanced by adding stronger leaf-springs and larger rear wheels. Only rarely did these vehicles resemble each other in detail, even though significant numbers were used by the Wehrmacht and allied forces. These converted vehicles were not used in combat roles, rather they saw service at bases in Germany and with rear-echelon units. On occasion they served as "Siren Cars."

In 1937 Opel in Rüsselheim attempted to interest the authorities in a four-door Kübelwagen on the chassis of the 1.3-liter delivery van. The vehicle was intended for use by police units or para-military fire brigades. Only one of these vehicles was built, since the Wehrmacht was more interested in light trucks with cross-country capabilities.

Left: Prototype of an Opel P 4 Tourer as an open four-door.

Right: An example of a Kübelwagen (military utility vehicle of similar concept to the Jeep) based on the Opel 1.2-liter box-type delivery van; the wheels of the Opel Super 6 have been mounted on the rear axle.

Commercial 1.3-liter vans were used by the Reich Propaganda Office as loudspeaker vehicles or talking film trucks, while Opel sedans and cabriolets (1.5-liter Olympia and 2.5-liter Kapitän) were preferred by war correspondents. Since the company assigned photographers to accompany Opel vehicles, today we have a large choice of interesting photographs to select from. This applies in particular to Kapitän-39 cabriolets, which were assigned to commanders and headquarters, and the 3.6-liter Admiral, which was used by senior commanders. At the same time, Admiral limousines were converted into small transports and tractors for light anti-aircraft guns. There were also ambulances built on the Admiral chassis. In most cases, however, these former limousines were overloaded and overtaxed in hard service at the front. What was more, the 3.6-liter machines consumed a great deal of fuel and towards the end of the war could no longer be driven. The 6-cylinder, 3.6-liter engines found new use in the Blitz 3-ton truck, a vehicle using

Opel 1.3-liter delivery van as a "talking film car" of the Reich Propaganda Office.

Below: A "Super 6" Kübelwagen with tool box in the rear.

Right above: Normal four-seat cabriolets were also converted into Kübelwagen by removing the two doors, without changing the standard rear end. The photo shows an Opel 2-liter Kübel in freezing rain in the Crimea.

Right: A Super 6 Kübelwagen which, judging by the large bullet holes, has already undergone its "baptism of fire."

the same engine, but had a much greater load capacity. This interchangeability of parts and even whole assemblies accounted for the great affection felt for Opel vehicles by the vehicle repair units.

Privately-owned Opel sedans were hastily painted in dull military colors at the Army Motor Transport Park.

Requisitioned civilian vehicles which have been assigned to the Luftwaffe, including a 1-ton Blitz owned by a country bakery, wait in a large garage to be put to use.

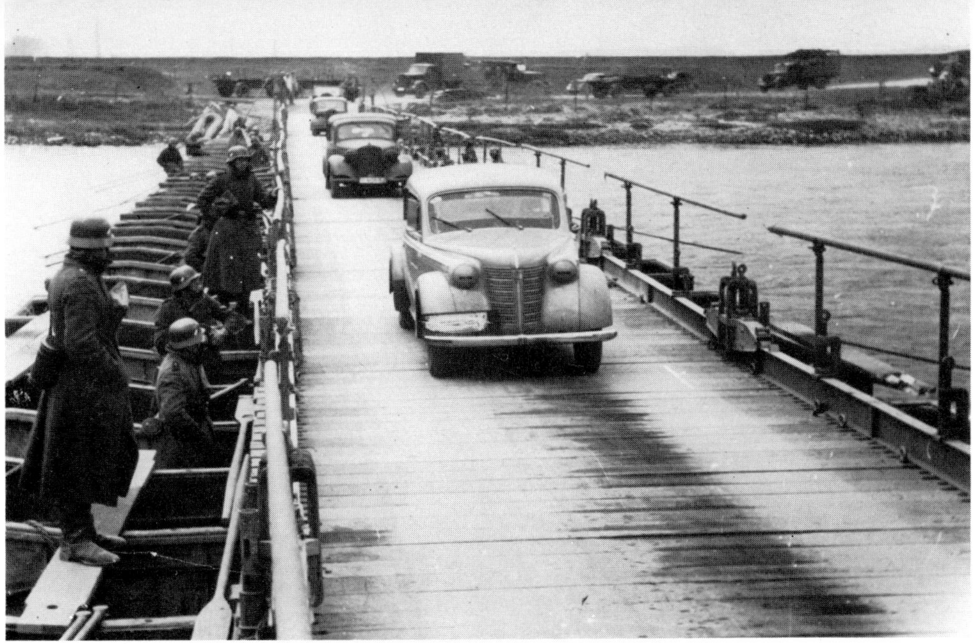

Above: The Germans were able to accelerate their advance by requisitioning Dutch civilian automobiles. The photo shows an Olympia 38 which, filled with soldiers, is crossing a pontoon bridge in May 1940.

Below: In Austria, too, Opels were "conscripted" and given military markings.

Above: A 1937 1.3-liter Olympia during a pause in the French Campaign.

Below: The Olympia Cabriolet was especially well suited for the role of column command vehicle.

Russian roads often resembled washboards; however, the Opel synchronous suspension handled them amazingly well.

Often, when the roads had become softened, the advance in Russia was only possible along railway embankments, as here in November 1941.

Above: The smaller Opel Kadett was rarely encountered in the Wehrmacht. The 1.1-liter was tough, however, and could, as seen here, be repaired with simple tools carried in the car.

Below: A war correspondent's Kapitän cabriolet crossing an anti-tank ditch on the isthmus of Perekop in 1942, watched by a Rumanian patrol. In the foreground is a Russian machine-gun.

Above: The Luftwaffe gladly made use of the Kapitän, as here during the advance in France in the spring of 1940.

Below: Press photographers also travelled about in the Opel Kapitän. The red chevron on the vehicle's licence plate indicates that the owner has special authorization to operate a motor vehicle in spite of fuel rationing. Only about 15 percent of all motor vehicles were so authorized (by decree of 6 September 1939).

Above: The last Opel Kapitän was built in October 1940, soon after the celebrations marking the completion of the one-millionth Opel automobile. There was little to celebrate, however: at the same time the manufacture of automobiles was terminated in favor of truck production.

Right: Opel had to advertise even in wartime. An advertisement from a 1940 issue of *Motor und Sport* shows Opel vehicles in action. The caption reads: "OPEL, the dependable one, able to master any terrain, equal to any challenge."

Reich Party Day 1938: The Adam Opel AG placed a number of its large Admiral cabriolets at the disposal of the diplomatic corps. It was the first time that the Admiral, which had been in production since May 1938, was used in an official function alongside the usual government Mercedes limousines.

Right: The 3.6-liter Admiral was issued only to the highest-ranking officers. Some of these vehicles were already in service with the military before the outbreak of war. Whenever a senior headquarters changed its location, several of the large Opel cars were always to be seen along with the more numerous Mercedes 540 K staff cars.

Below: The fighter aces Mölders (above) and Galland also used the Admiral as a service vehicle.

The Admiral's powerful 3.6-liter engine and stable chassis resulted in a very capable light truck. One advantage lay in the interchangeability of parts between the Admiral and Blitz-series vehicles.

Below: Ambulance aircraft conveyed the wounded to hospitals in Germany for special treatment. Here patients are transferred from the "White Ju" to an Admiral ambulance.

Opel cars and trucks remained in use throughout the war. They proved to be outstandingly simple, robust and dependable, which explains the great affection felt for them by the troops. Many field-gray Opels also took part in official events, as here at the funeral of *Generalfeldmarschall* Rommel in the summer of 1944, alongside vehicles by Horch, BMW and Mercedes.

An advertisement in a 1941 issue of *Motor und Sport*. The caption reads: "Ready for hard use day or night, equal to the greatest demands. OPEL, the dependable one."

OPEL BLITZ FOR THE WEHRMACHT

Despite the absence of specialized military versions, large numbers of Opel vehicles were used by the Wehrmacht on account of their proven unitized construction and the interchangeability of replacement parts among different vehicle types. More than 100,000 Opel Blitz trucks saw service as supply and troop transports, radio or field repair-shop trucks, ambulances and buses. This was a figure reached by no other make used by the Wehrmacht.

Since the German automobile industry was turning out only 1,000 vehicles per month at the beginning of 1940, increasing numbers of Blitz trucks from private industry were pressed into service by the Wehrmacht. As a result, civilian Opel 3-ton trucks with special-purpose bodies were to be found in all theatres alongside the 1.5-ton and 3-ton "S" trucks and later the all-wheel drive Blitz, which were more suited to front-line duties. The majority of the civilian vehicles were box-type delivery vans formerly used by furniture movers, or Opel Blitz tour buses. While, in many cases, the requisitioned Blitz trucks received the standard Wehrmacht flat-bed or box body and could only be distinguished from standard military trucks by the different style of driver's cab, most of the buses remained unchanged. They did, however, lose their large areas of chrome trim and bright paint schemes.

Likewise unchanged were the 1-ton Opel Blitz commercial vans used in limited numbers by the Wehrmacht. These served mainly as courier vehicles or as carriers of loudspeakers for the Reich Propaganda Office. The light trucks were also employed for the delivery of newspapers and magazines as well as mail to the armed forces. With a low unit weight of 1,340 kg this smallest Blitz could be expected to carry over 1,050 kg. Because of this a few 1-ton Blitz trucks of the types "2.0-12" and "5200" were used as tractors for the 2-cm Flak 30 after the outbreak of war. The Type "5200" 1-ton Blitz entered production in 1938 with the installation of the 37 h.p., 1.5-liter short-stroke engine from the Olympia automobile.

The appearance of the Opel Blitz 2- and 2.5-ton trucks in 1935 made available to industry a faster truck with a greater load capacity. As a dump truck this type of vehicle played a major role in the construction of the Westwall (the line of fortifications on Germany's western frontier).

In later years limited numbers of early pre-1936 versions of the Blitz such as these were still to be seen in service with the Wehrmacht.

From 1938 Opel also offered a Blitz with a 1.5-ton pay load, which was powered by the new 55 h.p., 2.5-liter engine used in the Super 6 and Kapitän. Although the small 1.5-ton Blitz (type designation 2.5-32) demonstrated a good performance and was equipped with lengthwise bench seats for the Wehrmacht, only 5,000 examples were built for the military before production ceased in 1942, as the military had little interest in the 1.5-ton class of truck.

The greatest production figures were achieved by the Opel Blitz 3-ton S, with 82,356 units produced with the 3.6-liter engine from April 1937 to the beginning of August 1944. The 6-cylinder engine produced 75 h.p., which was later limited to 68 h.p. at 3,000 r.p.m. off roads.

This dependable, light 3-ton truck enjoyed great popularity among the units. The simple, yet robust Opel design proved itself in the confusion of war and was superior to many specially-designed vehicles. Problems which could not be repaired on the spot were usually taken care of by the relatively efficient spares service, as the factories in Rüsselheim and Brandenburg were running at full speed. Another important factor in its success was the interchangeability of parts among Opel vehicles. Many unit leaders took great pains to have a "purebred" Opel fleet, employing all manner of compensation to achieve this. Similar methods sometimes had to be used to obtain replacement parts. During the French Campaign, for example, a bottle of Cointreau was often exchanged for an Opel transmission.

Replacement parts for Daimler and BMW vehicles frequently had to be obtained by soldiers on leave in Germany. The only similar bottleneck in spares for Opels concerned the supply of new radiators, without which the vehicles could not operate.

Quite often it happened that a Blitz, once it had been started with the aid of blow-torches or hot water following the coldest of Russian nights, was called upon to help start all the other vehicles by towing them. But it was not only for these reasons that the Opel Blitz was preferred as a transport vehicle in the Army's panzer regiments. Opel vehicles all burned *Otto* fuel (gasoline), which was more readily available than diesel fuel. Moreover, the experiences of the units with the Standard Diesel Truck had been less than favorable. "If something went wrong with them — they were often finished!" The soldiers were equally critical of the Mercedes 3-ton diesel, which was not suited to the hardships of the

The smallest Opel Blitz was the 1-ton Type 2.0-12 which, from 1934, was equipped with the engineering of the 2-liter, six-cylinder Opel automobile. Its use by the armed forces was for the most part limited to the Reich Propaganda Office which employed the vehicle as a loudspeaker-equipped van. The Blitz 1-ton was also used as a lighter, faster transport for military mail and the armed forces newspaper. After war broke out a few examples were used as tractors for the 20mm Flak 30, the 1-ton Blitz serving alongside the Krupp L2H with the first army flak units (for example in MG-Btl.46). With 36 h.p. and a piston displacement of 2 liters, the small Blitz was a robust and dependable vehicle.

Left: Opel Blitz 1-ton platform truck as an "advertising vehicle" for the 1936/37 *Winterhilfswerk* (Winter Relief Program, an annual charity drive).

Russian war.

Despite its dual rear wheels, the 3-ton Blitz was superior to many other 3-ton trucks on account of its low ground pressure, which was a result of the

Above: With a ground clearance of 209 mm, drivers had to give a wide berth to deep holes on roads such as these.

vehicle's low weight. Blitz trucks often got through soft terrain which had defeated the Büssing-NAG 4.5-ton truck, with its two powered axles. If the Blitz did bog down in soft ground, tracked vehicles had to assist. Much was expected of the *Maultier* (Mule) half-track truck, which was soon to be delivered to the units.

There were also important tasks at the front for the "Propaganda Blitz." Here a loudspeaker van is helped across a plank road.

This Blitz 1-ton shows the strain of the French Campaign. German troops reached Montdidier before the cease-fire with Germany and Italy went into effect at 01.35 on 25 June 1940.

In the years from 1935 the 1-ton Blitz came to be used as a personnel transport, especially by police and fire departments. This special body, designed by Magirus, was also suitable for the carriage of ladders and fire-fighting equipment.

The 1-ton Opel Blitz had been in production since 1934 with the 2-liter, six-cylinder engine. In 1936 all Blitz trucks received a new driver's cab in order to permit the installation of new engines in the 3-ton truck beginning in 1937 and the small Blitz trucks in 1938. The 1-ton received the 4-cylinder engine from the Opel Olympia, while the 1.5-ton was equipped with the 6-cylinder engine from the Kapitän.

With a pay load of 950 kg, the 4-cylinder Blitz with the model designation "5200" entered production in 1938. It was dropped from production in 1940 in favor of the 3-ton Blitz. The "P" on the front mudguard indicates a vehicle in service with the Ministry of Propaganda. In the foreground is an 1937 model Opel Kadett cabriolet.

Driver's cab of a 1-ton Blitz with tachograph. The foot-operated starter is plainly visible above the gas pedal.

6 October 1939: a speech by the *Führer* is broadcast by this Blitz of the District Film Depot of the NSDAP (Nazi Party).

The latest news was also broadcast in foreign combat zones. The photo shows an Opel Blitz 1-ton on a French National Highway on 10 June 1940.

With the outbreak of war the 1-ton Opel received minor modifications for service in the field; the vehicle in the photograph below has a blackout driving light and carries "field" equipment.

Right: Through clever use of parts already in production, Opel was in a position to develop a especially light and inexpensive truck with a pay load of 1,500 kg. Its very low weight allowed the vehicle to operate off roads, prompting the Wehrmacht to order the Opel Blitz 1.5-ton with a wide variety of body types. The "small" Blitz was especially well suited for the ambulance role with single tires in the rear and softer leaf springs. As well as being an ambulance, the 1.5-ton also saw service with the Luftwaffe fire-fighting service as the Kfz. 345.

Below right: In the Wehrmacht the 1.5-ton pay load class was of little significance. Nevertheless, the Opel Blitz 1.5-ton was the most widely used light truck. The last 1.5-ton was delivered to the Wehrmacht in November 1942, after 16,410 examples had left the production line.

Left above: Opel produced over 16,400 1.5-ton Blitz trucks. Approximately 10,000 were delivered to the Wehrmacht and served there as small personnel transports or special purpose vehicles. The Blitz also saw service as a tractor for light anti-aircraft guns or the 37mm Pak (anti-tank gun).

Above: Ambulance on Opel Blitz 1.5-ton chassis.

Left: The 1.5-ton Blitz during a rest stop. This version of the Blitz was to be the first Opel vehicle to go back into production after Germany's surrender.

Below: The few Blitz trucks built with special coachwork, as here with a Pullman limousine body, were used as staff vehicles or served senior officers as command vehicles.

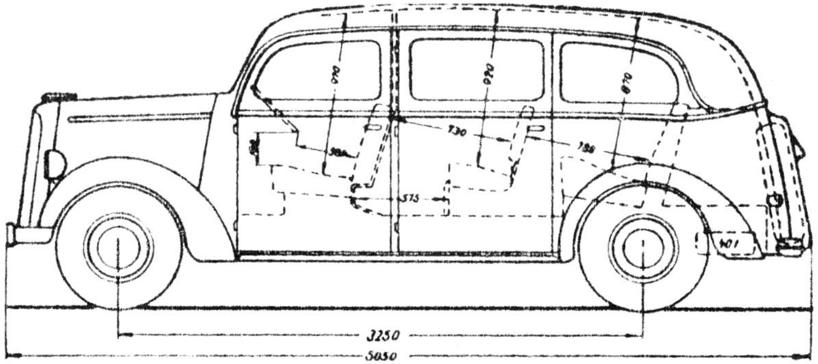

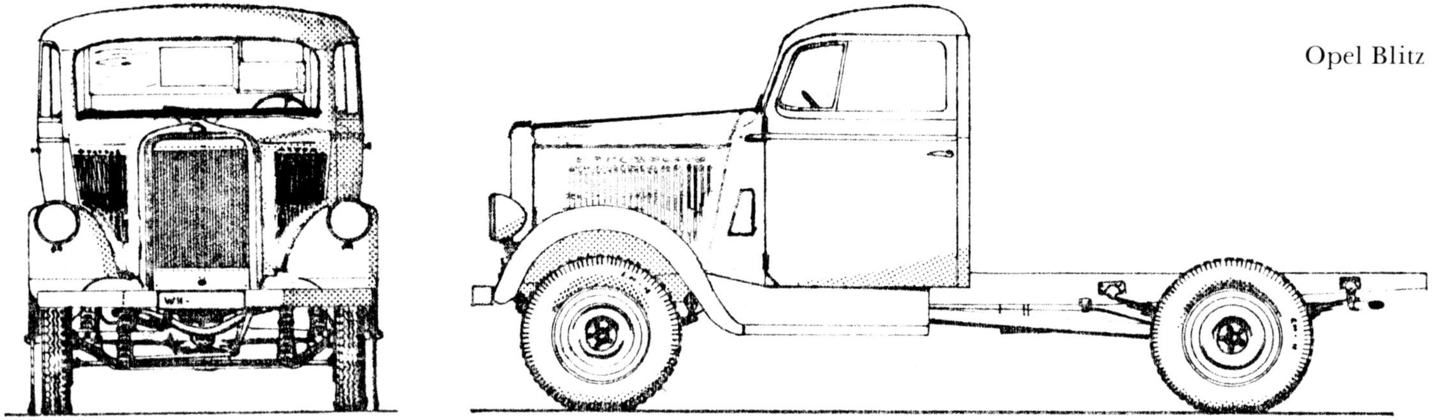

Opel Blitz 1-ton, 1936 Model

Opel Blitz 1.5-ton, 1938 Model

Opel Blitz 3-ton "S", built from 1936 to 1944.

Opel supplied the 3-ton Blitz to the Wehrmacht with various standard body types. By the end of September 1940, 31,674 Blitz trucks had been delivered to the Wehrmacht.

Right: Handover of a row of 3-ton Blitz trucks to the ***Leibstandarte SS Adolf Hitler*** in the giant motor vehicle hangar.

Below: Opel Blitz 3-ton trucks were employed on every front, in the West as well as in Russia.

Opel Blitz trucks carrying supplies rolled in endless columns, sometimes cross-country, as here in France on 12 June 1940.

Advance in the West.

Below: Destroyed houses in the Belgian border city of Tournai in May 1940.

Below: When the bridges had been blown by the enemy, emergency bridges were built, as here over the Meuse near Sedan on 22 May 1940.

Opel vehicles were also used in Norway. Here one is seen taking captured Norwegian soldiers to the rear.

Below: The "wire layers" of the signals units also made use of the light and maneuverable Blitz 3-ton.

05.00 on the morning of 31 September 1939: German troops in Blitz trucks cross the Narew River near Zambrow in Poland.

The "wire layers" of the signals units also made good use of the light and maneuverable Blitz 3-ton.

The advance in Russia made great demands on men and machines. The signals units with their heavy radio trailers provided a clear demonstration of what could be expected of a standard 3-ton Opel.

Below: When the ferries over the Dniepr near Cherson could not run because of pack ice in the river, the supply columns had to make a detour of 200 km. The heavy staff omnibus frequently became stuck.

Below: A knocked-out Soviet armored car along the side of the road near Alupka in the Crimea in April 1942.

Radio station on an Opel Blitz 3-ton in Fedorovka, west of Taganrog, in front of a typical Ukrainian farm house. In April 1942 straw matting was used as camouflage as well as protection for the engine against the cold. The body of the truck has been propped up with stones in order to relieve the pressure of the heavy superstructure on the leaf-springs and to prevent swinging when someone entered the radio station.

After the engine had been warmed by fire, blowtorch or hot water, the truck was placed in gear and rocked back and forth in order to free the pistons and bearings which had become resinified in the cold.

Rommel favored the Opel 3-ton truck for his ***Afrika Korps***. The legend goes that columns of Blitz trucks were sent to raise dust in the desert, simulating the tanks which Rommel was lacking and inducing the British to retreat.

On the move through the Driana Oasis near Benghazi.

Below: Benghazi Harbor. German engineers built a loading ramp over a grounded steamer, over which two ships could be unloaded simultaneously. Large numbers of Opel Blitz 3-ton trucks were employed to haul away the goods.

From 1943, vehicles in Germany had to be driven by wood-gas generators. Opel therefore delivered the 3-ton Blitz with a wood-gas generator, the Imbert installation proving to be the best.

Below: A Mercedes L 701 licence-built Opel which survived the war. The licence-built 3-ton trucks all had the standard driver's cab and carried no markings.

Right: Cadets fueling an He 111 at the Heinkel factory airfield at Marienehe.

Below: Opel Blitz 3-ton as an aviation fuel tank truck (Kfz. 385). The open driver's cab on this vehicle may not have been a variation of the early standard cab, but a special one designed to suit the needs of an airfield tank truck, as was the case with the earlier Mercedes LG 3000 (Kfz.384). Vehicles of this type also served as alcohol tank trucks (*B-Stoffwagen*) with V2 rocket batteries. Many Luftwaffe units did not have specialized tank trucks, but rather were equipped with standard Opel 3-ton trucks with a flat-bed-mounted fuel tank and pump.

Below: Fuel is siphoned from a French freight train captured in June 1940.

THE OPEL MAULTIER

Designated the Type "3.6-36S/SSM", the Opel Blitz half-track vehicle was a special version of the Blitz 3-ton S. The "Maultier," which entered production in October 1942, featured a rear-axle-driven tracked running gear, which considerably improved traction in difficult terrain. This was intended to allow the vehicle to carry a full load over the soft, muddy roads encountered primarily in Russia.

The half-track cargo truck, the result of a demand by the Waffen-SS, was built in considerable numbers. In the 2-ton pay load class Ford produced 13,952 vehicles, Opel approximately 4,000 and Magirus about 2,500. Daimler-Benz built a 4.5-ton model, of which 1,480 examples were produced. About 300 Opel Maultier chassis served as carriers for the Sd.Kfz.4/1, a 15-cm *Panzerwerfer* 42 (Sf.) (self-propelled 15-cm rocket launcher). Opel had designed its own tracked running gear which possessed superior self-cleaning characteristics. This was passed over, however, in favor of a copied version of the British Carden-Lloyd tracked running gear. While the Opel running gear allowed a maximum speed of only 38 kph, the Ford version with the Carden-Lloyd running gear achieved a speed of 50 kph in high gear. The Wehrmacht, however, prescribed a speed of 20 kph in order to increase the life of the running gear.

The mounting of the running gear on the 3.6-S was relatively simple. A few additional drilled holes were necessary to mount the unit, and a shorter drive-shaft drove the normal rear axle, which had been moved forward, to which were bolted the drive sprockets. A supplementary braking system served to steer the case-hardened manganese tracks, which had been taken unaltered from a Panzer I. The three different Maultier models were compared in Test Report IIId 4/43:

During trials in 1943 Opel engineers demonstrated how easily a standard 3-ton S-Blitz could be transformed into a powerful Maultier with the help of the Opel-designed tracked running gear.

COMPARISON OF THE THREE MAULTIER VERSIONS

		Dimension	Maultier Opel	Maultier SS	Maultier LC (Ford)
Gesamtgew.	für die	kg	6.600	6.600	6.950
Nutzlast	Erprobung	kg	2.900	2.850	2.900
Achsdruck	vorn bel.	kg	1.000	1.000	1.420
Achsdruck	hinten bel.	kg	5.600	5.600	4.740
Länge		m	6,02	6,02	6,25
Breite		m	2,24	2,24	2,29
Höhe		m	2,50	2,50	2,18
Bodenfreiheit	hinten	mm	320	320	275
Leistung		PS	68	68	95
Leistung je t Gesamtgew.		PS/t	10,3	10,3	13,7
Kraftstoffverb.	Straße	1/100 km	53	46,5	56,5
Kraftstoffverb.	Waldrdstr. Kdorf	1/100 km	97	96	106
Höchstgeschw. bei Nenndrehzahl	kleinst. Gg,	km/h	4,8	4,8	6
Höchstgeschw. bei Nenndrehzahl	größter Gg.	km/h	38	38	50
Kettenauflegelänge		mm	1.715	1.800	1.865
Spez. Bodendruck		kg/cm^2	0,69	0,6	0,53
Lenkbremse Art			Fußhebel hydr.	Handhebel mech.	Handhebel mech.
Wendekreisdurchmesser mit Lenkbremse		m	13,2	13,3	16,0
Wendekreisdurchmesser ohne Lenkbremse		m	14,5	15,9	22,0
Vorderrad bis Mitte Kettenauflage		m	3,120	2,950	4,013

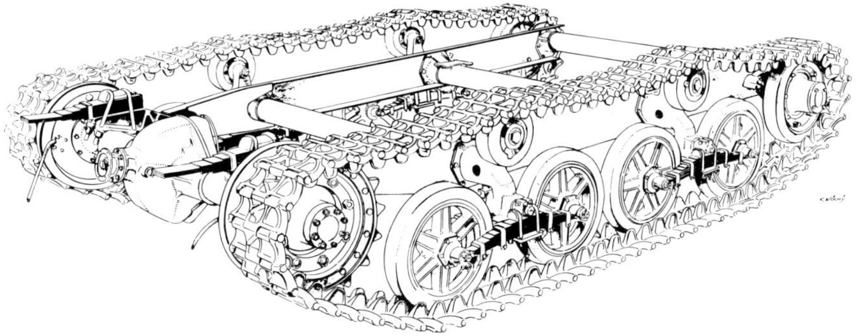

The Opel tracked running gear was not used on production Maultier vehicles, even though it was simpler to maintain and had better self-cleaning characteristics.

Photographs of the Opel Maultier in action are rare, even though about 4,000 examples were delivered. Here several Opel half-track trucks are seen on the Eastern Front where the German withdrawal has already begun.

Below: Opel Maultier chassis as carrier for the *Sonderkfz.* 4/1, a 15-cm *Panzerwerfer* 42 (armored rocket launcher). (Drawings by Molin/Austria)

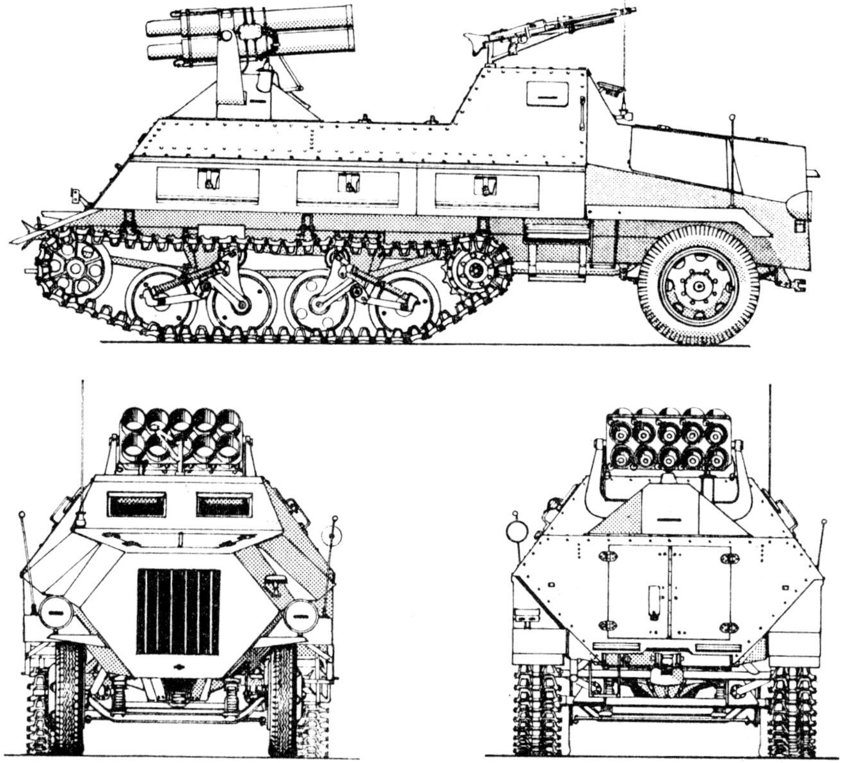

THE ALL-WHEEL DRIVE BLITZ

As early as 1938 the Wehrmacht's favorable experience with the standard Opel 3-ton Type "S" truck led the Plenipotentiary of Motor Vehicles, *Generalmajor* Schell, to encourage Opel to produce a version with the simplest possible form of all-wheel drive. With its low weight and great robustness, it was expected that an all-wheel drive Blitz would achieve new levels of cross-country capability and versatility. Development, which began immediately, resulted in a 3-ton all-wheel drive truck which could climb a 70% gradient with a full load, as compared to a 30% gradient by the standard rear-wheel drive version.

The new vehicle's wheel base of 3,450 mm was 150 mm shorter than the standard S-Type as a result of the rear axle being moved closer to the driver's cab. This gave the Blitz 3-ton "A" its characteristic chunky shape. Other features that easily distinguished it from the S-Blitz were the sloped corners of the doors over the rear of the front fenders and the engine hood, whose lower edge ran in a straight line above the front fenders.

Technically, the A-Type differed in having a transfer case and forward driving axle with differential. All-wheel drive could be engaged or disengaged with a single movement of an auxiliary shift lever, which significantly enhanced the vehicle's flexibility. With the availability of rear-wheel or all-wheel drive, the driver had a total of ten forward gears and two reverse gears at his disposal.

The all-wheel drive Blitz, which had been in production since 1940, was revealed to the public for the first time at the Vienna Spring Exhibition of 1941. Although over 25,000 all-wheel drive Blitz trucks were built before the destruction of the production facilities in Brandenburg — a figure unequalled by any other German all-wheel drive truck — it was never possible to supply the fighting troops with enough of these popular vehicles. Restricted supplies of vital raw materials

Above: The chassis of the Opel Blitz 3-ton all-wheel drive truck, more than 25,000 examples of which were built.

Below: The all-wheel drive Blitz as delivered from the factory, fully equipped.

limited the production of transfer cases and front axles. As a result, a maximum of five of these cross-country-capable trucks was assigned to each motorized regiment. Eventually, in 1943, the employment of the Blitz 3-ton A was specified in an order of the day: only vitally important supply vehicles, such as field kitchens, ambulances and radio trucks were to employ the all-wheel chassis. Small numbers of these modernized radio-truck bodies on all-wheel drive Blitz chassis were produced by the Ludewig firm in Essen, which manufactured the standard Wehrmacht bus body.

At the same time as the new all-wheel Blitz was entering service, Opel was incorporating im-

A 3-ton all-wheel drive Blitz which survived the war in Upper Bavaria and today is in the caring hands of a military collector. Likewise the NSU Kettenrad with its Opel Olympia engine. (Photo: S. Maier)

A photograph of the factory-installed radio station in an Opel Blitz.

Below: Because more Blitz 3-ton all-wheel drive trucks were demanded by the armed forces than German industry could deliver, the cross-country capable trucks were issued only to selected units. Among these were the signals units, who preferred to use the all-wheel drive Blitz as a radio truck.

Flooded roads, such as here in Russia in 1942, were no problem for the all-wheel drive Blitz.

provements based on the lessons of experience into series production. For example, from 1942 all three-ton trucks were delivered with an additional leaf spring on the rear axle to cope with the constant high loads experienced at the front.

The final models of the Opel Blitz featured a "material saving driver's compartment" whose rear wall was made of wood. This resulted in a visibly higher roof line and a slightly more vertical windshield. The wooden standard driver's cab specified for other trucks was never installed on Opel produced vehicles. It was, however, used on examples built under licence by Daimler-Benz and was planned for the version scheduled to be built by Borgward in Bremen.

This plan did not come to fruition, however, on account of the massive Allied bombing campaign against the German armaments industry. The Opel truck factory on the Havel was also a target of an Allied bombing raid. On 6 August 1944 several British Liberator bombers destroyed the entire factory installation in a precision daylight attack delivered from an altitude of 8,000 meters. Reconstruction of the Brandenburg facility, which had been producing 2,500 trucks per month, was impossible. Nevertheless, Opel Blitz trucks continued to be delivered to the fighting forces from Mercedes licence production. In June 1942, in the face of considerable opposition, the Daimler-Benz works in Mannheim had ceased production of its own Mercedes Type L 3000 S, which had proved unsuitable for front-line use, in favor of a licence-produced version of the Blitz 3-ton. Designated the Mercedes L 701, it carried no markings or manufacturer's nameplate. Externally, it was distinguishable only by the angular standard driver's cab. This Opel truck was the only 3-ton Opel of which production continued after the war, albeit with the original driver's cab (1946 to 1950 in Mannheim, then by Opel until 1954 with 467 examples produced).

A Blitz 3-ton all-wheel drive truck in Africa.

Below: An all-wheel drive Blitz at Benghazi harbor.

The Blitz Bus

From the beginning of Blitz production the program included, in addition to the 2-ton and 2.5-ton trucks with 3,400-mm and 4,000-mm wheelbases respectively, an extra-long 2.5-ton frame with a 4,650-mm wheelbase which was intended as a base for passenger bus bodies for 28 to 30 persons. With the introduction of the new 1938 Model the company's share of the passenger bus market increased from the 25% of 1936 to 39%. The vehicle's high cruising speed was partly responsible for its success, but the main reason was its favorable payload-to-weight ratio. Due to the modern weight-saving construction methods used, a 37-seat Blitz bus weighed only 3,500 kg, or 95 kg for each passenger. This inspired many coachwork factories to select the Opel chassis for installation of their streamlined metal bodies.

The Opel bus was also taken into service by the military, which lacked the capacity to rapidly transport troops. Requisitioned from their proud owners, they soon lost their bright chrome trim. Most of the buses did not survive their military service.

Conventional bodies were built for city and overland bus services, especially for trailer towing operations, in which the Opel Blitz could be expected to haul a total of ten tons. The overwhelming majority of these vehicles were used by the German Post Office and State Railway. In the final years of the war all new replacement vehicles were fitted with bodies with integrated wood-gas generator installations, which were built into the rear of the vehicle.

Wehrmacht buses, the majority of which had bodies built and installed on Opel frames by Ludewig, served as transports for military personnel and the wounded, as well as command and communications centers. The military bus was a simplified version of the standard commercial Opel Blitz passenger bus. From 1939 to 1944 Ludewig Bros. of Essen delivered over 2,880 of these military buses with the standard body to the Wehrmacht. While, until 1943, all of the bus bodies were metal-skinned, from 1944 some were built with fiberboard panels. These Wehrmacht buses, which were equipped with wooden benches for 30 soldiers, had a radius of action of 280 kilometers on roads. However, the Opel bus was not suited only to the rapid movement of personnel. As the Model W39 these spacious vehicles also served as command and map vehicles, as mobile orderly offices and were also used by the medical services. There were also a few Opel Blitz versions that had box-shaped bodies mounted on their low-frame chassis. These were employed as workshop vehicles or "printing and type-setting vehicles for propaganda purposes."

The older Opel Blitz 2.5-ton bus was rarely seen in use by the Wehrmacht as a troop carrier. This "conscripted" Model 1935 bus is seen accompanying the advance in France in the early summer of 1940.

The 3-ton Blitz with low-frame chassis was the ideal base for bus and repair-shop truck bodies. The Opel factory delivered these chassis to many coachwork factories, including F.K.F., Voll, Kässbohrer and Ludewig.

Immediately after being requisitioned, the bright two-tone finish of the Döbelner Coachwork Factory (Saxony) was replaced by a coat of field-gray. This former tour bus was used by the Todt Organization.

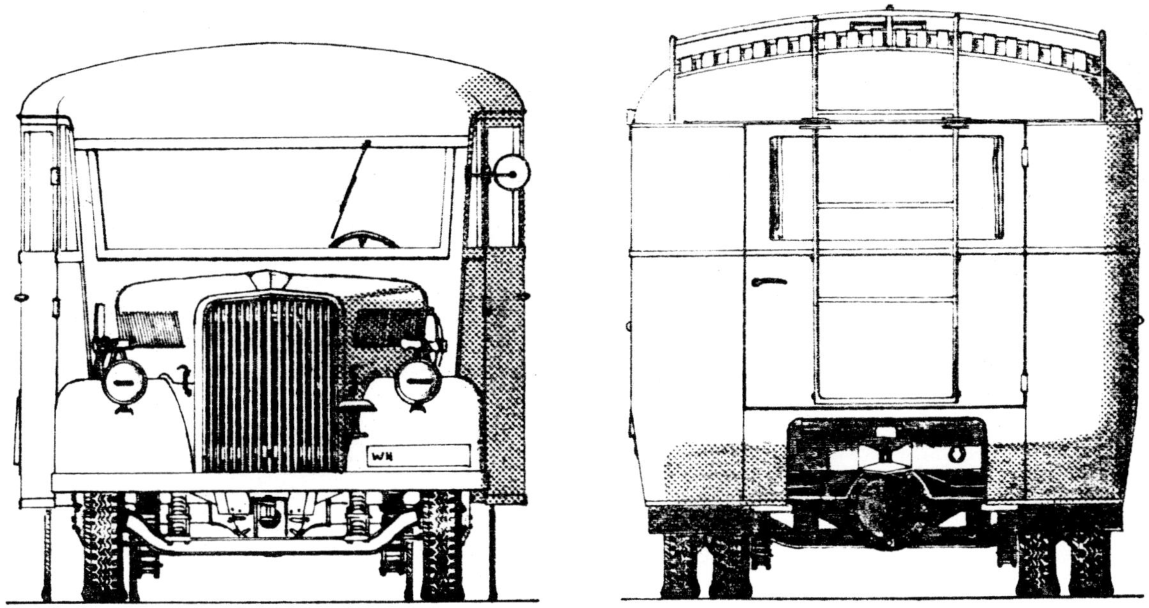

The Opel bus with standard Wehrmacht body was manufactured in large numbers by the Ludewig Coachwork Factory in Essen.

Machine-gun mount close to the driver's seat in an Opel bus.

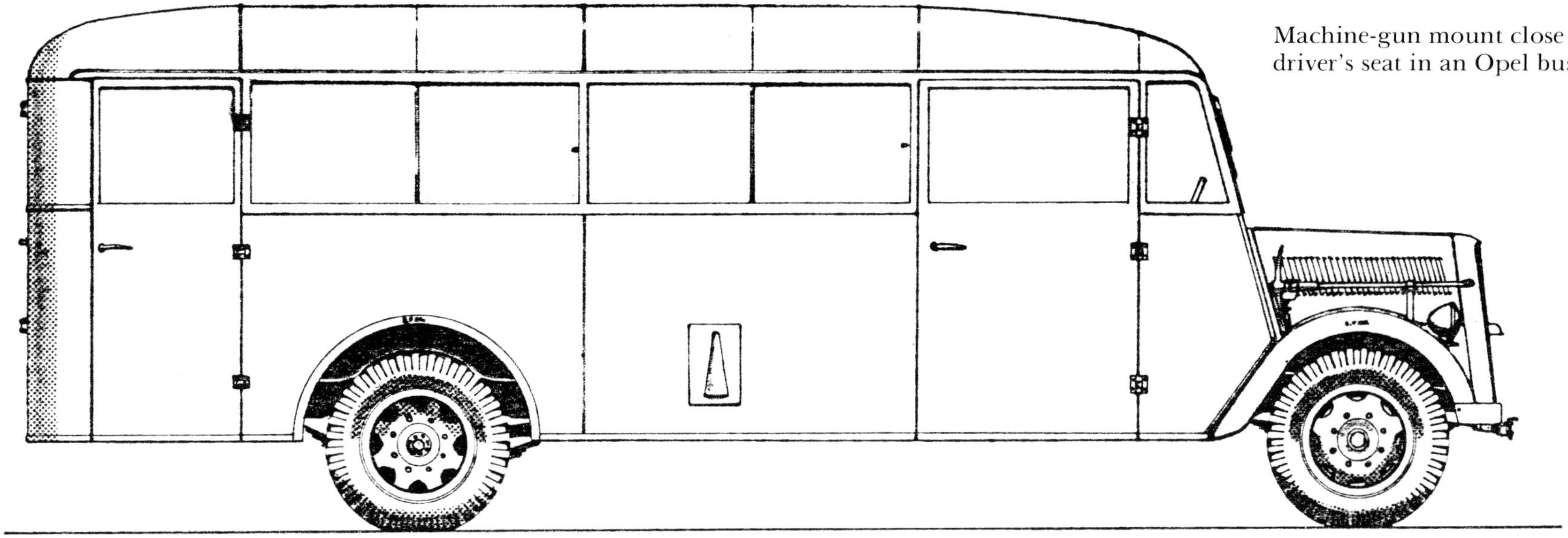

Opel buses with the new-style streamlined body were also requisitioned by the Wehrmacht, where they were robbed of their gleaming chrome parts.

In 1944, because of fuel shortages in Germany, vehicles had to be fitted with wood-gas generator drives. Kässbohrer delivered several Generator-Omnibuses on Opel 3-ton low-frame chassis for use by the German Post Office and the German State Railway. The "gas factory" was housed in the rear of the all-steel buses.

A former line-bus accompanies a column in Poland in October 1939.

Opel Blitz Statistics

By production year:

Modelle	1,1 ltr. 1,3 ltr. Geschäfts-wagen	1 to	1,75 to 1,5 to	2,5 to 3,0 to	Gesamt
1931	1.348	–	677	2.039	4.064
1932	1.294	–	–	1.792	3.086
1933	1.191	62	–	2.236	3.489
1934	1.492	4.038	–	2.950	8.480
1935	3.458	4.289	–	5.222	12.969
1936	7.496	5.427	–	8.833	21.756
1937	6.837	6.446	–	10.891	24.174
1938	4.499	1.361	5.577	14.859	26.296
1939	3.335	2.935	5.066	16.485	27.821
1940	410	379	1.678	17.605	20.072
1941	–	–	3.485	15.947	19.432
1942	–	–	604	18.262	18.866
1943	–	–	–	23.232	23.232
1944	–	–	–	16.146	16.146
1945	–	–	–	–	–

By production model:

Modell	Typ	Bauzeit	Stückzahl
P4 1,3 ltr Geschäftswagen	1396	11. 1935 – 02. 1938	15.672
P4 1,1 ltr Geschäftswagen	6100	03. 1938 – 04. 1940	7.405
Blitz 1 to 6 Zylinder	2,0-12	12. 1933 – 03. 1938	21.437
Blitz 1 to 4 Zylinder	5200	10. 1938 – 06. 1940	3.500
Blitz 1,5 to	2,5-32	01. 1938 – 11. 1942	16.410
Blitz 2 to/2,5 to, 3,5 ltr	3,5-34/83	12. 1930 – 03. 1937	21.196
Blitz 3 to, 3,5 ltr	3,5-36+47	05. 1936 – 07. 1937	5.605
Blitz 3 to + 3 to "S"*	3,6-36	04. 1937 – 07. 1944	82.356
Blitz 3 to lang	3,6-42	06. 1937 – 07. 1944	14.122
Blitz 3 to Niederrahmen	3,6-47	05. 1937 – 07. 1944	8.336
Blitz 3 to Allrad	3,6-67	07. 1940 – 07. 1944	24.981
Blitz 3 to/3,6 ltr Gesamt		04. 1937 – 08. 1944	129.795

*includes Blitz 3-t S/SSM "Maultier" half-track vehicles and Panzerwerfer on Blitz "S" chassis.